Kids Nonfiction -
662.88 HARDY
Hardyman, Robyn
Biomass energy
33410018215683   09-15-2022

T5-COA-155

DISCARDED
Valparaiso-Porter County
Library System

Energy Evolutions

BIOMASS ENERGY

Robyn Hardyman

CHERITON
CHILDREN'S BOOKS

Please visit our website, www.cheritonchildrensbooks.com to see more of our high-quality books.

First Edition

Published in 2022 by **Cheriton Children's Books**
PO Box 7258, Bridgnorth WV16 9ET, UK

© 2022 Cheriton Children's Books

Author: Robyn Hardyman
Designer: Paul Myerscough
Editor: Victoria Garrard
Proofreader: Wendy Scavuzzo
Picture Researcher: Rachel Blount
Consultant: David Hawksett, BSc

Picture credits: Cover: Shutterstock/Konstantin Romanov; Inside: p1: Shutterstock/Rudmer Zwerver; p3: Shutterstock; p4: Shutterstock/Alsem; p5: Shutterstock/Apixelstudio; p6: Shutterstock/Oliveromg; p7: Shutterstock/Mr. Prawet Thadthiam; p8: Shutterstock/Smileus; p9: Shutterstock/Nostal6ie; p10 Shutterstock/Holger Kirk; p11: Shutterstock/Rudmer Zwerver; p12: Shutterstock/Bertold Werkmann; p13: Shutterstock/Dugdax; p14: Shutterstock/Dan Lewis; p15: Shutterstock/David Tadevosian; p16: Shutterstock/Oleksiy Mark; p17: Shutterstock/Northstars; p18: Shutterstock/Roschetzky Photography; p19: Shutterstock/Antsipava Volha; p20: Shutterstock/Ryan Fletcher; p21: Wikimedia Commons/Spartan7W; p22: Shutterstock; p23: Shutterstock/Frontpage; p24; Shutterstock/Monkey Business Images; p25: Shutterstock/Belish; p26: Shutterstock/Celio Messias Silva; p27: Shutterstock/Reisegraf.ch; p28: Shutterstock/Thatsanaphong Chanwarin; p29: Shutterstock/Chung Toan Co; p30: Shutterstock; p31: Shutterstock/RikoBest; p32: Shutterstock/Miles Mortimer; p33: Shutterstock/Hywit Dimyadi; p34: Shutterstock/Perception7; p35: Shutterstock/Shine Nucha; p36: Shutterstock/Jerry Lin; p37: Shutterstock/Ye.Maltsev; p38: Shutterstock/Rob Crandall; p39: Shutterstock/Maliwan Prangpan; p40: bio-bean; p41: Shutterstock/Aedka Studio; p42: Shutterstock/EQRoy; p43: Shutterstock/Tom B Paye; p44: Shutterstock/Belish; p45: U.S. Navy photo by Adam Skoczylas.

All rights reserved. No part of this book may be reproduced in any form without permission from the publisher, except by reviewer.

Printed in the United States of America

Contents

Chapter 1 What Is Biomass Energy?.................4

Chapter 2 Wood and Waste Biomass.................8

Chapter 3 Liquid Biofuels.................16

Chapter 4 Where in the World?.................22

Chapter 5 New Ideas.................32

Chapter 6 Biomass Energy Today.................42

Glossary.................46

Find Out More.................47

Index and About the Author.................48

CHAPTER 1
What Is Biomass Energy?

Biomass is any material that comes from either plants or animals. It includes grass clippings, vegetable peelings, sawdust, animal dung, and animal fat. This material contains energy that has come from the sun. Plants store the sun's energy in their roots, stems, and leaves when they use sunlight to grow, and that energy is passed on to animals when they eat the plants. This energy can be released from both plants and animals, and it is called biomass energy.

Cleaner Energy

Biomass energy is part of the solution to our need for **cleaner energy**. At the moment, more than half of the energy the world uses comes from burning coal, oil, and natural gas in **power plants**. These energy sources are called **fossil fuels**. We extract fossil fuels from the ground, but there is a limited supply of them. That is why these energy sources are called **nonrenewable**. It took millions of years for them to form underground, so we cannot replace them. Fossil fuels are burned in power plants to make electricity, and this process releases harmful gases into the air that trap the sun's heat and warm our **atmosphere**.

The energy from the sun is stored by plants and passed on to the animals that eat them. Both plants and animals can be used to create biomass energy.

4

This is causing changes to the **climate** around the world, and this can have catastrophic effects on life on Earth. Scientists tell us we must urgently reduce our use of fossil fuels. Biomass is a **renewable** source of energy, because new plants can be grown. However, biomass energy is not completely unlimited like other types of renewable energy, such as the wind or sunshine.

Energy from Living Things

We release the biomass energy stored in plant and animal material by burning it. The heat energy this releases can be used directly to heat water or buildings, or it can be used to **generate** electricity in a power plant. Biomass can also be converted into liquid fuels, called biofuels, and a gas called biogas. These fuels are then burned to release their energy.

BIG Issues
Carbon Dioxide

Burning biomass to create energy releases carbon dioxide into the atmosphere. This is a harmful **greenhouse gas** that is contributing to **climate change**. However, plants are also very useful stores of carbon dioxide. Throughout their lives, they soak carbon dioxide up from the air and use it to grow. Some people say that this cancels out the harm done by burning the plants, making biomass energy **carbon neutral**. Others say that we cannot afford to add carbon dioxide to the atmosphere and plants are more useful as carbon stores.

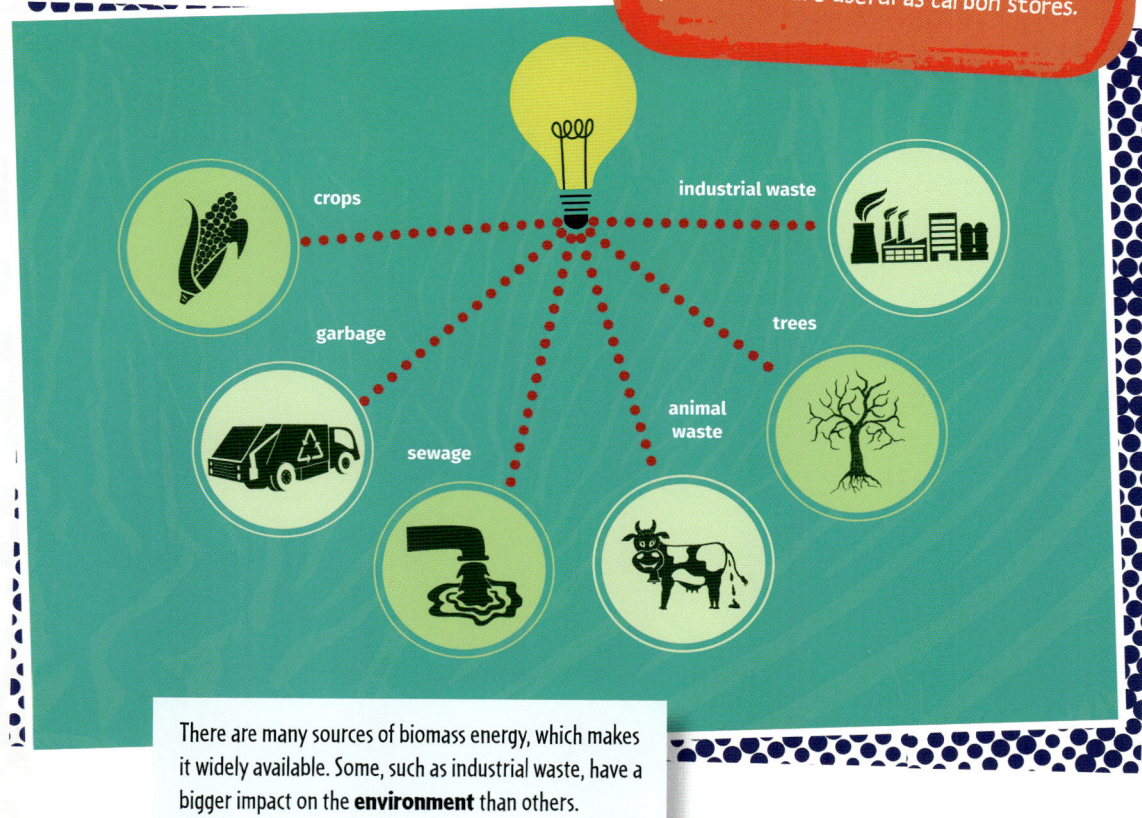

- crops
- industrial waste
- garbage
- trees
- sewage
- animal waste

There are many sources of biomass energy, which makes it widely available. Some, such as industrial waste, have a bigger impact on the **environment** than others.

How Does It Work?

The sun's energy, stored in plants and animals, is what gives us biomass energy. That energy can be released in several different ways. Some of these ways are very simple and we have been using them for thousands of years. Others are new and cutting edge. Today, more and more technologies are being developed to help create as much energy as possible from the natural resources around us.

Burning Biomass

One way to release the energy in biomass is to burn it. This is an old technique. People have been burning wood for hundreds of thousands of years, since humans first discovered fire. We have used wood fires to give us heat and for cooking our food.

We have also been burning fats such as beeswax and oil for thousands of years to give us light. The modern biomass industry involves cleaner and more complex systems for extracting as much energy as possible from these resources to supply the largest number of people.

We use biomass energy when we burn wood on a campfire.

Burning biomass creates heat. This heat can be used directly to warm water and homes. It can also be used indirectly to create steam. The steam is then used to power a **generator** to create electricity. We can burn new materials, such as wood, for biomass energy, or we can burn waste products, such as our everyday trash. Wood is a popular choice in places with a lot of forested land, such as in Scandinavia. In places where there is not a lot of wood to spare, burning trash is a more popular option.

Cows release a lot of harmful methane gas into the atmosphere in their belches. However, their waste can be used to create biogas, which is used as fuel.

Making New Fuels

Another way to create energy is to use the biomass to make new fuels. For example, some crops that contain a lot of sugar are used to create ethanol, a type of alcohol that can be added to gasoline. Biodiesel can be made from vegetable oils and animal fats. It can be used in vehicles or as a heating oil. Biomass can be used to create fuel in the form of a gas. This is made from animal and human waste. The waste is taken to a treatment plant, where the gas it gives off is collected in containers and made ready for use.

CHAPTER 2
Wood and Waste Biomass

Wood is one of the most important sources of biomass energy. In countries where forests are plentiful, it offers valuable opportunities for providing local and sustainable energy. There are several different ways in which it can work.

Woody Waste

There is a lot of leftover material when forest plantations, or areas of trees that are planted, are harvested. This includes the tree tops and branches that are left lying on the ground once the tree trunks have been cut down. If the leftover material is used nearby as an energy source, it can help people make money and help the environment. Processing tree logs into regular sizes, such as timber planks, also produces a significant amount of waste wood at the sawmill. In fact, up to half of a log can be left over as waste in the form of sawdust, bark, shavings, and wood chips. This waste does not have to be thrown away, however. It can be burned at the sawmill and the heat used for processes

Wood biomass can include any part of the tree, from the leaves and branches to the waste left over at the sawmill.

on the site, such as drying timber. This is an excellent way to generate clean, sustainable, and cheap energy, and you cannot get much more local than this. One sawmill in Victoria, Australia, is saving thousands of dollars a year by burning its waste wood in a wood-fired boiler and using the heat for the **kilns** that dry the timber. The excess steam from this process is also used to make electricity, saving the business even more money.

Heat for Growers

Waste wood chips can also be used to heat farmers' greenhouses. Although the initial cost of installing the wood-fired boiler is quite high, the fuel is so much cheaper than the alternatives that those costs are saved after two or three years, and then the business continues to save a huge amount every year.

Pellet Power

Waste wood can also be processed, or turned, into pellets, which are then sold as fuel. They can be made using wood from anywhere, including waste wood from towns, cities, and construction sites. This is one of the main ways in which wood is used for making biomass energy.

BIG Issues: Saving Money

Using wood waste to create energy requires less money if it can be done locally. If the waste has to be taken a long way to a power plant, the high cost of transportation makes it less worthwhile. Power plants that rely on wood waste for energy also need a constant supply, which is why locally sourced wood is a better option.

This biomass power plant is powered by burning wood chips.

9

Waste Biomass

As we consume more and more, we are creating more and more garbage. Fortunately, there is a way that we can turn our wastefulness into a benefit, and that is by using the trash to release energy.

Food Waste Win

To turn food waste into energy, the waste is enclosed in a large container. The oxygen in the air inside is removed, creating what is called an **anaerobic** atmosphere. The **bacteria** that occur naturally in the waste start to digest it, or break it down. As a result of this process, several products are released: carbon dioxide, water, a dark slurry called digestate, and methane. Methane is a gas that can then be burned to generate electricity or **refined** to use as a fuel.

When local authorities build anaerobic digester plants, they are creating a neat, closed-loop system that works financially and environmentally. Local waste is used to make clean energy for local people. Another benefit is that the waste does not have to go to a huge dump in the ground called a landfill site. Smart changes in the technology are making it even more **efficient** all the time.

This biogas plant produces a lot of **fertilizer** as a **by-product**. It is used by farmers to add **nutrients** to their land.

Completing the Cycle

There is another huge benefit from this process. The digestate that is left over from the anaerobic digestion is full of nutrients that can be added to soil to help new plants grow. It is sold to farmers to be used as a fertilizer. This is a great system—using plant waste to grow more plants.

If the waste ends up in landfill sites, however, it can still be used for biomass energy. As it breaks down, it slowly releases methane gas, too. This can be collected and used to generate electricity.

This farm has installed a small plant for turning the waste from its cows into biogas. The farmer can make some extra income by selling this biogas, or use it to power processes on the farm.

Making Methane

Animals also create masses of waste from their digestion. Belches from cows produce an enormous amount of methane. Although this cannot be collected, the methane in the animals' waste can be. Dairy farms and pig farms can use anaerobic digesters to produce biogas from manure and used bedding material from barns. If they collect their animal waste in large ponds, they can cover the ponds and capture the biogas. They can use the gas to heat their water and buildings, or as fuel in a diesel-engine generator to create electricity.

One of the advantages of this type of biomass energy is that the supply of waste is constant and reliable, so producers can be sure their energy output will be, too. This is not always the case with renewable energy sources, such as solar and wind power.

Environmental Costs

Using the natural resources of wood and organic waste to produce energy makes good sense in so many ways, not just because it is replacing the use of fossil fuels. Some of the techniques used in these processes, however, have a negative impact on the environment. This means that we have to carefully weigh the positive and negative sides of this type of biomass energy.

Polluting the Air

The main drawback with burning wood waste and garbage is air **pollution**. Wood smoke contains harmful particles that can affect people's health. In developing countries, wood fires are the most common form of cooking and heating. Many people from those countries have breathing difficulties as a result of the particles in the smoke. Modern techniques for burning wood waste as biofuel are less polluting. There are also rules to protect air quality.

It is important that biogas power plants like this one capture the carbon dioxide they create so it does not enter the atmosphere.

Forests that are used for biomass energy must be carefully managed. That way they can continue their valuable role of absorbing carbon dioxide from the atmosphere.

Waste-to-energy plants can create harmful **pollutants**. The regulations, or rules, in the United States require these plants to use air pollution control devices. These include scrubbers, fabric **filters**, and other technology to capture the pollutants before they can be released into the air. The ash left over from burning waste at these plants also contains harmful chemicals, such as dyes and inks from textiles and packaging. The chemicals must be removed from the ash before it can be released from the plant. Some of the ash is then used to make concrete blocks and bricks.

Methane itself is a highly polluting greenhouse gas. It causes more damage in the atmosphere than carbon dioxide. Burning methane turns it into carbon dioxide and water vapor but because carbon dioxide is less harmful than methane, the overall impact is lower.

BIG Issues
Removing Trees?

Trees are very important to our planet. They take carbon dioxide out of the air and trap it in their branches and leaves for many years. When we remove trees, we are losing a valuable friend in the fight against rising levels of carbon dioxide in the atmosphere. Also, when we burn trees, we release the carbon dioxide they contain, making matters even worse.

Using waste wood, however, does not involve cutting down trees that would otherwise have been left to grow forever. In forests, the trees are carefully managed. They are harvested when they are fully grown and replaced with new trees. The waste from cutting down trees would be produced anyway, so it makes sense to use it to create clean, local energy, so we can burn fewer fossil fuels.

Using Biomass Energy

Around the world, discoveries in biomass energy are transforming the industry. As the need to reduce our use of fossil fuels is becoming more and more urgent, scientists and governments are looking at wood and waste biomass as a promising alternative. Innovation is happening across the industry, and some exciting new projects are ready to make the most of this valuable resource.

Setting a Trend

In New Hampshire, the research university Dartmouth College announced in early 2019 that it would invest $200 million into a biomass energy heating system to replace its existing central heating. This type of large-scale project sets a good example for others to follow. In the United Kingdom (UK), many new biogas plants supplied by food and animal waste have begun in recent years. In 2013, there were 100 anaerobic digestion plants in the UK, but by the end of 2018, that number had increased to more than 640.

Europe is also big on using wood waste, from fueling wood burners with wood chips to generating electricity. One world-leading project is at the massive Drax power plant in the UK. Drax was powered by coal, but

Dartmouth College in New Hampshire is leading the way by installing a new heating system powered by biomass.

The waste from chicken farms is a useful source of biomass energy.

it is switching to being powered entirely by wood pellets. In 2018, it had switched four of its six powering units to this new technology. Although burning pellets still produces some carbon dioxide, Drax is now working on new technology to capture that carbon dioxide before it is released into the air. This would make this biomass energy even better for the environment.

Help from Chickens

The litter from poultry farms is rich in nutrients and has been used as a fertilizer for many years. Litter is a mix of manure, water, feathers, and spilled feed, and it is now also being used for energy. This reduces the amount of waste the farm must get rid of, recovers the energy in it, and gives the farmer another source of energy or income. Darling Downs Fresh Eggs was the first egg producer in Australia and one of only a few in the world to power its business using renewable electricity generated from its litter. It uses an anaerobic digester to produce methane gas, which it uses for heating and creating electricity. The heat warms the sheds for young chicks and the water for washing the eggs. The company now uses about 60 percent less electricity from the **grid** than it did before switching to biomass.

CHAPTER 3
Liquid Biofuels

Another important way of creating energy from biomass is to use plant material to make new fuels called biofuels, which can be used in transportation. There are two main ways this is done. The first is to grow plants especially for making fuels, and the second is to use the residues, or leftovers, from harvested plants that are grown for other purposes, such as for food.

From Field to Pump

How can plants be turned into fuel? Plants that contain a lot of sugar can be used to make an alcohol called ethanol. Corn, sorghum, sugarcane, sugar beets, and barley can be used in this way. The ethanol is then blended with gasoline. Most gasoline in the United States today contains about 10 percent ethanol. Any vehicle can use fuel containing this amount of ethanol. If there is more ethanol in the fuel, the vehicle's engine must be specially adapted, or changed, for the fuel.

Biodiesel is another biofuel. It is made from vegetable oils and animal fats. Vegetable oil that has been used for cooking works well, so many restaurants and other food outlets sell their used oil to biofuel plants. Blended diesel usually contains about 20 percent biofuel, and it can be used in regular diesel engines. It produces lower levels of most air pollutants than regular diesel fuel.

Transporting biofuels by freight train is less polluting than putting the containers on large trucks.

Transportation is one of the biggest uses of energy in the developed world. It accounts for about one-third of all the energy used in the United States, for example. Transportation is one of the biggest polluters because petroleum-based fuels such as gasoline spew harmful gases into the atmosphere when they are burned in vehicle engines. This is why it is important that we find cleaner fuels. In 2019, biofuels accounted for just 5 percent of the fuels consumed in transportation in the United States. Ethanol had a 4 percent share, and biodiesel had 1 percent.

Using Up Leftovers

When food crops are harvested, there is always material left over as waste. This may be sugarcane husks, fruit skins, grape seeds, nut shells, or straw from a grain crop. All these materials can be used to make biomass energy. At the moment, experts are working on smart new ways to make the process better, so that the costs are earned back by releasing as much energy from the crop residues as possible.

On this farm, corn is being specially grown and processed to make ethanol. The ethanol is added to gasoline.

Grown for Energy

The value of plants as sources of energy means that some farmers now grow crops especially for that purpose. They are used to make ethanol and biodiesel for transportation. Although this means that vehicles produce fewer harmful emissions, people are concerned about the impact on the environment and on the production of food.

Power Crops

In some places, power crops are grown in huge quantities. The most common crops grown for this purpose are corn, sorghum, and barley, because they have a high sugar content. Corn currently provides most of the liquid fuel from biomass in the United States, for example. The problem with growing power crops is that they require a lot of natural resources, such as water. They also take up land that could be used to plant crops for food. For these reasons, it makes more sense to grow crops that are **native** to a region, such as trees and grasses. These do not require management and do not use up other resources. Another sustainable solution is to grow "cover crops." These are extra crops that are grown on the land over winter, after the food crops have been harvested. The cover crops are grown to add nutrients to the soil and are then harvested for energy.

Growing sorghum for biomass has become big business, but people now accept that it is not a sustainable way to produce energy.

18

Cover crops grown in winter do not compete with food crops for land, and they are a more sustainable form of biomass energy.

Fast-Growing Trees

Some trees that grow quickly are suitable for energy use, such as willows and poplars. As old trees are cut down, new trees are planted. Other trees will grow back very quickly if they are cut back hard in a process called **coppicing**. This allows the trees to be harvested every three to eight years, for 20 to 30 years, before they need replanting. They can grow as much as 40 feet (12 m) high in the years between harvests, providing plenty of biomass material. Willow, poplar, and maple trees can be grown for biomass energy in the northern United States, while sycamore, sweet gum, and eucalyptus are grown in the South.

BIG Issues
Efficient Energy?

Using large areas of land to grow crops for fuel is not an efficient or sustainable way to produce energy. The crops require large amounts of fertilizer and **pesticide** to grow. The production of these creates harmful greenhouse gases. More energy is actually used in the growing and harvesting of these crops than can be produced from them as biofuel. They also use up the precious resources of water and land. This is not an efficient use of farmland, which could be better used to produce the food that people need.

Biofuels in the Air

Aviation, or air travel, is one of the biggest sources of greenhouse gas emissions. We fly goods from one side of the world to the other, and millions of people travel the world for business and leisure every day. We urgently need to find ways to make this form of transportation cleaner, and biofuels have an important role to play.

Future Fuels

The aviation industry is committed to using more biofuels, promising that half of all international aviation fuel will be biofuel by 2050. This presents a big challenge, however, because it will use an unacceptably huge amount of land to grow crops such as palm oil to make these biofuels. New sources need to be found.

Industrial Gases

In 2018, the airline Virgin Atlantic flew the first-ever **commercial** flight from Orlando, Florida, to London in the UK, powered by biofuel made from industrial waste gases that were converted into ethanol. This new fuel formed only 5 percent of fuel blend for the flight, but the plan is for it to be blended 50/50 with standard jet fuel in the

Virgin Atlantic airline is working with researchers and businesses to use more biofuel in its airplanes.

future. This would reduce the fuel's harmful emissions by around 65 percent. The fuel is being developed by a company called LanzaTech. The company hopes to produce enough fuel to power all of Virgin Atlantic's outbound flights from the UK.

LanzaTech produces its advanced fuel by recycling industrial waste gases, such as those produced from steelmaking and other heavy industrial processes. The super-smart alcohol-to-jet process used to make the fuel used in the Virgin Atlantic flight was developed with the help of Pacific Northwest National Laboratory and the United States Department of Energy (DOE).

Energy from Mustard

Also in 2018, a Qantas airplane powered partly by mustard seeds became the world's first biofuel flight between Australia and the United States. The 15-hour flight used a blended fuel that was 10 percent derived from an industrial type of mustard seed.

This crop does not replace food crops, because it is grown after the food crop is harvested, when the land would otherwise not be in use. Experts say that compared with jet fuel, this biofuel reduces emissions by 80 percent over the fuel's life cycle.

BIG Issues
Working Together

Scientists and the aviation industry need to work together to solve the problem of sustainable jet fuels. In 2018, a biofuels specialist company called Neste signed an agreement with aviation fuel giant Air BP, to work together to develop and distribute new sustainable jet fuels. They, too, would be biofuels that do not involve massive land use, because they are made from waste products.

Air BP has a well-established global network for distributing jet fuel at airports.

CHAPTER 4
Where in the World?

Around the world, some countries are using biomass energy more than others. Biomass currently represents around 14 percent of all the world's energy supply, and with innovations in the industry, this figure looks likely to rise in the coming years.

Electricity Evolution

Around 70 percent of the electricity that is produced from biomass today is being generated in North America and Europe. This is because countries there have large, sustainable forests that can supply a constant source of wood pellets to use in power plants. There are some biomass power plants in less developed countries, however, and these are valuable for delivering electricity to areas that are not on any grid. For example, Kenya gets about 75 percent of its total energy from biomass, and India gets around 50 percent. Biomass power plants in both countries are using agricultural waste to generate electricity.

This biomass energy project in Kenya is making biogas from water hyacinth plants.

22

Using Heat

Most of the biomass energy around the world today is used in **rural** areas of developing countries, where half the world's population lives. There, the biomass is often used in open wood fires for cooking and heating.

In industrialized countries, biomass energy is being used to provide the heat for some processes. A relatively low level of heat can be needed for sterilizing (or making bacteria-free), washing, cooking, drying, and many more processes. Renewable energy sources are perfect for delivering this type of heat, as well as for preheating before another heating source is used. Because it takes a relatively large amount of energy to raise the temperature of water, even a moderate amount of preheating can reduce the need for fossil fuels—and save money.

Transportation Solutions

The International Civil Aviation Organization (ICAO) wants to see a big increase in the amount of biofuel used in aviation across the world. In 2017, 25 countries rejected the ICAO's proposal to set a high target for this, because it would mean turning over vast areas of land, including a lot of farmland, to growing palm oil. This would threaten the livelihoods of millions of people in the developing world. Since then, experts have suggested that there are other ways to make biofuels for aviation, such as using waste gases from industry instead.

For vehicles on land, the United States leads the world in the amount of ethanol it uses in its gasoline. Brazil turns sugarcane into ethanol, and some cars there even run on pure ethanol. In Europe, biodiesel made using palm oil is more common.

Soybeans produce more energy than corn, but growing them uses valuable land. This truckload of soybeans is being delivered to a power plant in Brazil.

The United States

The three main sources of biomass in the United States are wood, garbage, and biofuels from plant crops, such as corn. All the biomass sources together contributed about 5 percent of all the country's energy in 2017, so there is still a long way to go for biomass energy to make a significant contribution.

Burning Wood

Wood is burned as logs and as pellets in homes across the country for heating. In 2015, it was the main heating source for about 3.5 million households. Much more wood and wood waste is used in industry and for generating electricity in power plants. As well as this, the wood pellet industry is a big export business. Innovative companies in the United States are processing huge amounts of wood into pellets, which they are selling to Europe and elsewhere to be used instead of coal in power plants.

Waste to Energy

Garbage is widely used at waste-to-energy plants and at landfill sites in the United States. There are more than 2,000 anaerobic digestion plants in operation across the country. In 2015, about 53 percent of the country's garbage was taken to landfill sites.

Many homes in the United States use wood as biomass to provide heating, either on open fires or in wood stoves.

This huge container of waste biomass will be processed to release its energy.

About 26 percent was **recycled**, and about 13 percent was burned as an energy source to make heat or electricity. The United States Energy Information Administration (EIA) estimates that in 2019, about 270 billion cubic feet (7.6 billion cu m) of landfill gas was collected at about 370 landfill sites. It was burned to create electricity.

Using Biofuels

Most of the ethanol fuel used in the United States is made from corn, but scientists are working on new ways to make ethanol from all parts of plants and trees, rather than just the grain. Most trucks, buses, and tractors use diesel, and the consumption of biodiesel has risen dramatically in recent years, from about 10 million gallons (7.8 million l) in 2001 to about 1.8 billion gallons (6.8 billion l) in 2019. The United States now uses more than 20 percent of all the biodiesel in the world.

BIG Issues
A Helping Hand

In their early years of development, renewable energy industries need government support to help with the risks and costs involved in creating new forms of energy. In the United States, the government has helped support the growth in biomass energy. The Renewable Fuel Standard Program, for example, was created by Congress to set binding rules for replacing fossil fuels with biofuels, to reduce greenhouse gas emissions.

Energy in Latin America

The countries of Central and South America have long been dependent on oil for their energy needs. Now, however, with high oil prices and the pressure to reduce their output of harmful greenhouse gases, they are looking to renewable energy sources, including biomass.

Biomass in Brazil

Brazil, the largest country in the region, has massively increased its use of biofuels in transportation. The country has always grown sugarcane, but now this sugarcane is used to make biofuel as well as sugar. In fact, Brazil has one of the world's most advanced green transportation programs, and ethanol must be added to all gasoline. The country has set itself a target of having 18 percent of all its energy needs met by biofuels by 2030. It is even thinking of exporting sugarcane to other countries in the region for their energy needs.

This biomass power plant in Brazil uses sugarcane to make ethanol.

Advances in Argentina

Argentina has a renewable energy program called RenovAr, with 147 projects across the country. Of these projects, 18 are biomass projects, and 36 are biogas projects. One new power plant is being developed by a US-based developer working with Argentina's largest renewable energy generation company. This innovative power plant is working on ways to get the maximum use from its sugarcane.

One by-product of creating ethanol from sugarcane is a substance called vinasse. The Argentinian power plant will be able to use the vinasse, too, to make fuel. This will allow the power plant to make more ethanol from the same amount of biomass. The sugar and ethanol industries in Argentina employ thousands of workers, so this is great news for the economy as well as the environment.

Chile's Cool Future

Chile imports, or buys and brings in, almost all of the fossil fuels that it uses for most of its energy generation. The government there accepts the urgent need to invest in renewables. A new law has been passed stating that 25 percent of its energy must be from renewable sources by 2025. The country has a lot of agriculture and forestry, so both of these could be used for biomass energy. In the south of the country, for example, there are large plantations of eucalyptus. The wood waste from harvesting these trees is perfect biomass material. This growing industry could bring new jobs to rural areas and introduce new ways to make money from agriculture and forestry.

Timber from eucalyptus trees is used in Chile as biomass for creating energy.

Energy in Asia

Many countries in Asia have traditionally had to import their gasoline and diesel from other countries. Now they are looking at the possibilities of biomass and developing smart new solutions to meet their energy needs within their own countries.

Changes in Thailand

In Southeast Asia, Thailand has a lot of potential for biomass. Experts think biomass could meet at least 15 percent of the country's energy needs. The biomass industry is now expanding quickly there, with the rice, sugar, palm oil, and wood-related industries producing plenty of crop residues. These are being burned to produce steam for generators in at least 10 large-scale operations across the country. Thailand also has many pig farms, and waste from these is used to create biogas. Biomass and biogas are becoming major energy sources in Thailand, supported by a government that offers tax **incentives** to encourage biomass businesses.

The number of biogas plants, like this one in Thailand, is increasing rapidly.

Japan and China

Biomass is taking off in the Far East, too. Japan aims to meet 22 to 24 percent of its electricity from renewables by 2030, and it is developing new biomass plants to help meet this target. For example, in 2019, the company Toshiba signed an agreement with the city of Omuta to build a large new biomass plant to help increase renewables.

In China, the amount of electricity generated by biomass has increased tenfold in the last decade. The country plans to double this by 2030, making use of the huge amount of biomass material from its agricultural and household waste.

Increasing in India

In India, there are increasing numbers of biomass power plants. Some are providing valuable off-grid electricity to remote areas. Others are feeding electricity into the grid. In agricultural areas, the process of using crop residues is beginning to expand, although the supply is rather unreliable, because many of the farms are small and do not use a lot of machinery. Farmers are also starting to look at growing crops such as bamboo and napier grass especially for power generation on marginal land, or land that is not useful for farming. There is a huge potential for biomass energy in India, once the practical problems such as limited farm machinery are overcome.

The residues left over after the harvesting of this rice crop will be perfect for creating biomass energy.

BIG Issues: Ready Residue

Power plants need a reliable supply of whatever resource provides their energy. It can be difficult to guarantee regular supplies of crop residue material, because crops are harvested only at specific times of the year. The challenge is to find money-saving ways to store the residue material after harvesting so it is always available for use.

Biomass in Europe

Biomass energy is big in Europe, where there are large areas of managed forest producing wood, huge areas of agricultural land producing crop residues, and millions of people producing household waste. All these resources are being used creatively to meet the energy needs of Europeans.

Using Wood

From fueling wood stoves to generating electricity, wood has an important role across Europe. Different countries use it in different ways. Europe produces and consumes more wood pellets than any other part of the world. In Sweden, wood pellets power district heating systems, where whole areas of a city are heated through a single central plant that distributes heat to buildings. It is a similar story in Finland. The UK has used less wood in the past, but that is changing. In 2019, a major new power plant opened in Yorkshire, UK, powered entirely by locally sourced wood waste. It will produce enough power for 78,000 homes. The largest biomass plant in the world is expected to open in the UK soon. The Tees Renewable Energy Plant will use 2.7 million tons (2.5 million metric tons) of wood chips to produce enough electricity for 600,000 homes. Most of the wood will be imported from Europe and the United States.

In some European countries, large areas are covered by forests, so wood biomass is in plentiful supply.

This biogas plant in Germany is located in an agricultural area, so that it can use the surrounding biomass material.

Goodbye Coal, Hello Garbage

The UK's biggest power plant, Drax, is in the process of converting entirely from coal to wood for its electricity generation, and another UK plant is following their lead. The same thing is happening in France, where the government has announced that the Cordemais coal power plant will also convert to biomass. France has committed to ending all coal-fired electricity generation in the next few years.

Across Europe, more anaerobic digesters are also being built to make biogas from the millions of tons of garbage produced every year. There are more than 350 anaerobic digester plants in Europe, with about half of them in Germany.

Biofuels That Are Unsustainable

The negative impact on the land has meant that growing crops especially for biofuels is scaling back in Europe. In 2019, the European Union (EU) ruled that biodiesel made from palm oil is unsustainable because it involves the destruction of too many forests. This means that biodiesel made mostly in Indonesia and Malaysia will be banned from the EU. However, Europe still aims to have 14 percent of all transportation fuels in every EU country coming from renewable sources such as biofuels by 2030.

CHAPTER 5
New Ideas

The biomass energy industry is dynamic and creative. Scientists and engineers are constantly researching new ways to make energy from the plentiful natural resources around us. Whether it is generating biogas from garbage, ethanol from crop residues, or electricity from wood waste, every aspect of this forward-looking industry is under the microscope.

Catching the Carbon

When wood is burned to produce heat for electricity generation, it releases a lot of carbon dioxide into the atmosphere. This is carbon dioxide that the trees have captured from the air over their lifetime. Although this makes the process "carbon neutral," carbon dioxide is still a harmful greenhouse gas. Its emission should therefore be limited. The latest idea is to capture the carbon dioxide at the power plant before it is released into the air. In the UK in 2019, a company specializing in carbon capture secured £3.5 million ($4.5 million) of investment from BP, Drax, and others. It is developing a chemical-based system to remove carbon dioxide from power plants, steelworks, and cement factories. The technique is being tested at Drax and is the first trial of its type in Europe.

Drax Power Station in the UK is switching from using coal to biomass in the form of wood pellets. Drax is also trialing ways to capture the carbon dioxide before it is released into the atmosphere.

Coming Up with Cool Solutions

Many more anaerobic digesters and power plants are being built in the United States. It is essential that each part of the process works effectively. One problem in the past has been that the biogas produced in an anaerobic digester contains water. If this gets into the power plant, it reduces its ability to create electricity and causes engine damage. One smart new technique to overcome this problem is to remove the water in a process called dehumidification. The machine reduces the temperature of the gas, which makes almost all the water in it condense into liquid. This leaves a clean, dry biogas for the power generator.

Bagasse is the fibrous matter that remains after sugarcane is crushed to extract its juice. It is used as a biofuel, but storing it properly is a challenge.

BIG Issues
Storing Biomass

At harvest times, a large amount of biomass material is produced by crops such as sugarcane. This may need to be stored, so that it can be fed into the machinery for making ethanol at a later date. One problem with this is that the crop residue is likely to either rot or burst into flames if it is not stored properly. If it rots, it releases harmful methane gas. If it catches fire, it cannot be used. Strangely, wet sugarcane residue catches fire more easily than dry residue. Researchers at one university in Australia are using mathematics to calculate the temperature and moisture in different-sized stockpiles of sugarcane residue. This will help them figure out the ideal amount that can be stored safely.

Awesome Algae

As our need for cleaner transportation fuels grows every day, the race is on for new forms of fuel, and scientists are working hard to find them. One possible new source is an **organism** that we see around us in rivers and ponds, and that is algae.

When it comes to the potential to produce fuel, no biomass can match algae in the quantity or range of fuels that it can produce.

Green Mats of Energy

Algae is the name given to a collection of small organisms that grow in water. Algae forms huge mats across the surface of ponds and other water sources, and can grow in either salty or fresh water, almost anywhere that is sunny. It has no roots or leaves and feels slimy when you touch it. All algae needs to grow are water, sunlight, and carbon dioxide. These natural resources are freely available and can be used without damaging any natural **habitat** or taking up land that could be used for growing crops.

Squeezing Out the Oil

Algae was first explored as a possible source of fuel in 1978, when the cost of gasoline soared. The key to its use as a fuel is that it contains a lot of oil. In the processing plant, there are three ways to extract this oil. In the first method, it is simply pressed, just like olives are pressed for their oil. This releases up to 75 percent of the oil. The second method adds a chemical that releases more oil from the algae that is left over after the first pressing. The third method extracts all the oil.

The algae is heated, and carbon dioxide is added. This turns the algae completely into oil, but this process costs more and takes longer than the other two methods. The extracted oil is then refined to produce a biodiesel fuel.

Industrial Growing

How can we produce enough algae to meet our demand for biodiesel? If we grow algae outdoors, we rely on sunny weather to make it grow. The water also has to be kept at a constant temperature, which can be tricky. Biofuel companies are developing "closed-loop production," in which the algae is grown vertically in clear plastic bags exposed to sunlight on both sides. The algae gets more sunlight, so it grows more and is also protected from contamination. Other companies are growing algae indoors in large round drums with artificial lighting and ideal conditions. The algae can be harvested every day, yielding more fuel.

Since algae needs carbon dioxide to grow, algae biofuel manufacturing plants are being built close to other energy plants where carbon dioxide is produced. This is a perfect solution to capturing that harmful carbon dioxide before it reaches the air, and using it to create even more energy.

The potential of algae as a source of biofuel is one of the most exciting areas of research in the biomass energy industry today.

Amazing *Azolla*

The world of plants has produced an impressive source of biomass energy that can be used for making biofuel. It is called *Azolla*. *Azolla* is a fast-growing plant that is unusual because it does not need to grow in soil.

A Superplant

Azolla is an aquatic plant that grows in fresh water, rather than salty water. It is a type of fern, and it is one of the fastest-growing plants on the planet. It can grow so quickly because it grows in water that also contains a type of bacteria that naturally extracts nitrogen out of the air. Nitrogen is a powerful fertilizer, so the *Azolla* is fed very well and grows quickly. In fact, it grows so quickly that it can double in size in just two days. It can even do this in water as shallow as 1 inch (2.4 cm) deep. That is truly a superplant!

Azolla grows incredibly quickly, even in wastewater.

Azolla and algae like this can work together to make biofuels.

BIG Issues
Cutting Edge

If *Azolla* is one of the fastest-growing plants in the world, could this be the solution to all our transportation fuel needs? Research into this solution is new, and there are challenges in converting such an enormous amount of plant matter into biofuel in a way that is not too expensive. *Azolla* and algae seem to have a lot of potential for the future, though. Some researchers have even called *Azolla* "a green gold mine."

A Zillion Uses for *Azolla*

As it grows, *Azolla* absorbs masses of carbon dioxide from the air. It is a perfect way to capture carbon dioxide and keep it from contributing to global warming and climate change. Once it is harvested, *Azolla* can be used for many things. It can be processed into pellets of nutritious feed for livestock or into a biofertilizer to be spread over farmland to make new crops grow. It is already being widely used for this purpose by rice farmers in the Far East.

Azolla's role as a biomass energy source, however, has two parts. First, the sugars and other substances it contains can be processed to make a biofuel. Second, and even more amazingly, *Azolla* and the water it lives in contain a lot of nitrogen. Once the *Azolla* has been processed, this nitrogen-rich water can be used for another purpose: to fertilize algae that is being grown to make biofuel. This makes the production of biofuel from algae even more efficient.

Exciting Hydrogen

One of the most exciting developments in transportation energy at the moment is the fuel source hydrogen. Hydrogen is a gas, and using it to power vehicles does not produce any harmful emissions. Energy industry innovators think that biomass could be used to make hydrogen, to help us clean up our transportation systems.

Hydrogen Cleanup

Hydrogen is not burned in a car's engine. Instead, it is used in a **fuel cell** inside the car to make electricity, which powers an electric motor that drives the vehicle. Inside the fuel cell, hydrogen is combined with oxygen and, in the process, it produces water and electricity. This is super-clean energy because the only waste product is water. This water is so clean, you could drink it! The fuel cell is a little like a battery that never runs out, as long as there is a supply of hydrogen in the tank. This makes hydrogen cars more attractive than electric cars, which have a limited range. You can also fill the tank in minutes, instead of waiting hours for a battery to recharge.

This gas station in Washington, DC, already has hydrogen fuel pumps.

Corncobs like these can be used in **gasification** plants to make hydrogen.

Gasification Goals

Although hydrogen is so abundant in the world, most of it is locked in water—water is made of hydrogen and oxygen. Water is plentiful, but the hydrogen is very tightly bonded to the oxygen. Biomass contains hydrogen, too, but in a less tightly bonded form. The chemistry is complex. The process used to create hydrogen from biomass is called gasification. It involves using heat, steam, and oxygen to release hydrogen. Carbon dioxide and carbon monoxide are also released through this process.

Gasification plants are being built and trialed in several countries, so scientists can learn the best ways to make this technology work well. The DOE thinks that biomass gasification could be in use in the next few years to create hydrogen for vehicles in the next few years. A wide range of biomass materials can be used. These include grasses, corncobs, wood pellets and wood waste, and even treated sewage.

Gas Networks

Hydrogen produced in gasification plants has another valuable and exciting use. It can be blended with natural gas and used in the standard gas network that supplies all our homes and businesses. There are plants working on blending natural gas with 20 percent hydrogen gas. Because hydrogen does not produce any harmful emissions, when this blended gas is burned to heat buildings, it produces less carbon dioxide. This could be a simple way to reduce our greenhouse gas emissions quickly.

New Energy Sources

Developing biomass energy is a complicated business. Although we are surrounded by biomass, sometimes the technology to convert it to energy may be too complex or too expensive to compete with other renewable energy sources, such as wind, solar, or geothermal energy. This is why the industry is working so hard to make biomass energy work and to find new biomass energy sources all the time.

Better Biofuels

Making biofuels from crops grown over huge areas of land is not an efficient way to create energy. It leads to **deforestation** and the loss of land that could be used to grow food. Plants that grow on marginal land, where there are no forests and crops cannot be grown, are much more promising.

Scientists think that a better way to make biofuels is from grasses and the offcuts of coppiced trees. These contain more cellulose, which is the material that makes up the cell walls of the plant. Although commercial production of ethanol from cellulose is a new business, in recent years, up to 10 million gallons (7.9 million l) of the product have been produced.

Bio-bean is a forward-thinking company that uses coffee waste to make logs that can be burned.

Ghana is the second highest producer of cocoa in the world. The husks of cocoa pods could be used to generate electricity.

Creative Thinking

Other innovators are looking in completely different places for sources of biomass energy. One British innovator named Arthur Kay started looking in his local coffee shop. He noticed that his cold coffee had a thin, oily film on top of it, discovered that there is oil in coffee, and wondered if the energy in that oil could be released by burning it. He set up a company called bio-bean to explore the technology, and he found that he could convert coffee waste into pellets. These can be burned to warm ovens or even to heat buildings. Arthur also created fireplace "logs" made from his coffee waste.

Another green technology project is looking at generating electricity from discarded cocoa husks. The husk is the outer covering of the seedpod. A British university is working with cocoa farmers in Ghana, Africa, to see if their cocoa waste can be used to provide them with off-grid electricity. Every ton of cocoa beans that is harvested generates 10 tons (9.1 metric tons) of husks. These could be converted into valuable electricity for rural areas that currently have only 15 percent electricity coverage.

BIG Issues
Tough Cellulose

Although cellulose makes up so much of a plant's weight, it is more difficult and expensive to convert it to energy than it is for seeds or grains, because it is so tough. This makes it an expensive option.

CHAPTER 6
Biomass Energy Today

Biomass is the largest source of renewable energy in the world today, and the International Energy Agency's (IEA) current forecast for the years to 2023 is that it will remain the biggest. This is because it is used much more widely for heating and in transportation than other types of renewable energy. Solar and wind power, for example, are mainly used for generating electricity.

Biofuel Giant

The IEA thinks that modern bioenergy is the overlooked giant of the renewable energy field. It represents as much energy consumption as wind, solar, and all the other renewables combined. For it to continue growing, the right policies from governments, such as financial incentives for biomass energy and financial **penalties** for non-renewable polluters, will need to be put in place to support the industry. There are many encouraging examples of good practice around the world.

Seattle-Tacoma International Airport has recently announced that it will be the first airport in the world to be heated entirely by renewable biogas.

Committing to Biogas

Seattle-Tacoma is the largest airport in the Pacific Northwest of the United States. It is committed to using biogas heating. The airport invited companies to apply for the contract to supply its heating boilers, to encourage the development of the technology. By 2030, the airport plans to reduce its greenhouse gas emissions by 50 percent from its 2005 levels.

A Quickly Changing World

The world of biofuels is moving quickly. In the aviation industry, for example, pressure is growing on the airlines to reduce their harmful emissions. In 2019, the International Civil Aviation Organization (ICAO) launched a scheme called Carbon Offsetting and Reduction Scheme for International Aviation, or CORSIA.

More than 70 countries have volunteered to participate in the scheme, and plan to offset their carbon emissions by investing in renewable energy or energy efficiency projects elsewhere. A later part of the scheme, though, is for the airlines to make their operations less polluting by using biofuel. The goal is to cut aviation emissions by half by 2050. This type of international cooperation is the way forward in tackling these huge problems that no single country can solve on its own.

Some companies are rapidly embracing the potential that biofuels have to offer. In 2019, for example, Qantas Airlines agreed to buy 8 million gallons (30.3 million l) of aviation biofuel every year from a US company called SG Preston. This will be used on all its flights from Los Angeles to Australia, blended 50/50 with standard aviation fuel.

High-performance cars have been considered to be big polluters because they produce so few miles to the gallon of fuel. But they run much more efficiently on a fuel blend of 85 percent ethanol and 15 percent gasoline.

Evolutions of the Future

The future of biomass energy looks bright. Scientists and other industry experts are constantly pushing the boundaries of technology to make better use of the plant and animal material that is around us in such a plentiful supply. With such an emphasis on intense research and investigation, many new ideas are resulting in even more sustainable ways to power our world.

The world's growing population is creating ever-greater amounts of waste, so it makes sense to use it to create the energy we need.

Capturing Carbon

Waste-to-energy plants are growing in number, using our everyday garbage to make electricity. The methane that is released by the garbage is burned to create electricity, but carbon dioxide is also released into the air, which is harmful. The way forward is to capture that carbon before it is released. The latest technology can do just that, and then use it to produce other high-value products such as vehicle and aviation fuel. Carbon capture is definitely the way of the future.

A plane known as the "Green Growler" completed flight testing of a 100 percent advanced biofuel at Naval Air Station Patuxent River, in Maryland.

Transforming Transportation

Biomass has a huge role to play in transforming our transportation systems from fossil fuels to greener alternatives. Already, it is leading the way in this, and there is a great deal of effort being made around the world to take it further. In 2017, the United Nations launched below50. This international collaboration project works to develop and grow the technology and the market for fuels that produce at least 50 percent fewer carbon dioxide emissions than conventional fossil fuels. The project has three main hubs: North America, South America, and Australia.

We need to continue creating biofuels from crops such as corn and sugarcane for a short time to meet our ambitious targets, but in the long run, this method is not sustainable. The land biofuel crops are grown on is needed for growing food. Instead, the future lies in biofuels made with food crop residues and nonfoods such as grasses, trees, and algae. If we can find a way to grow and process enough algae in indoor, controlled conditions, this could meet a huge part of our fuel needs in the future. In Latin America, for example, where a lot of land is used for growing sugarcane for making ethanol, this would be an excellent step forward. The land currently used to grow sugarcane would be freed for growing food that people badly need.

BIG Issues
Working Together

To solve issues such as climate change, countries around the world must work together. The atmosphere does not recognize country borders, and we will all feel the effects of global warming. Thankfully, there are signs that the governments of enough countries want to work together to protect our future.

45

Glossary

anaerobic without oxygen being present
atmosphere the blanket of gases around Earth
bacteria microscopic organisms
by-product an extra product created by making the main product
carbon neutral resulting in no release of carbon dioxide into the atmosphere
cleaner energy energy that does not pollute the environment
climate the regular weather conditions of an area
climate change the changes in climate around the world caused by the gradual increase in the air temperature
commercial describes organizations and activities that are concerned with making money
coppicing cutting back trees to encourage them to regrow strongly
deforestation cutting down all the trees in an area to make the land available for other uses
efficient able to achieve maximum productivity with minimum wasted effort or expense
emissions something, usually harmful, that is put into the air
environment the natural world
fertilizer a material that is added to plants to provide nutrients essential to their growth
filters devices that remove solids from a liquid in a similar way to a sieve
fossil fuels energy sources in the ground, such as coal, oil, and natural gas, that are limited in quantity
fuel cell a device that produces electricity directly from a chemical reaction
gasification a process by which something is converted to a gaseous mixture
generate to make
generator a machine that converts energy into electricity
geothermal related to energy harnessed from the heat within Earth

greenhouse gas a harmful gas, such as carbon dioxide, that collects in Earth's atmosphere and traps the heat of the sun
grid the network that distributes electricity from power plants to consumers
habitat the home of a living thing
incentives payments or reductions that provide strong reasons for doing something
innovation a smart new way of doing something
kilns ovens for baking pottery or drying timber
native found naturally in a place
natural resources things we use, such as water and oil, that are found in the natural world
nonrenewable will eventually run out
nutrients parts of food that living organisms need to be healthy
organic related to the natural world, not human-made
organism a living thing such as an animal or a plant
penalties punishments or removal of something, such as money or privileges
pesticide a substance that is used to control pests, including weeds
pollutants harmful substances
pollution harmful substances in the environment
power plants places where energy is created
recycled used again after it has been thrown away
refined made purer
renewable describes energy created from sources that do not run out, such as light from the sun, wind, water, and the heat within Earth
rural related to the countryside
sustainable able to protect the environment by not using nonrenewable natural resources

Find Out More

Books

Krajnik, Elizabeth. *Biomass Energy: Harnessing the Power of Organic Matter* (Powered Up! A STEM Approach to Energy Sources). PowerKids Press, 2018.

Lowe, Alexander. *Renewable Energy in Infographics*. Cherry Lake Publishing, 2020.

Martin, Cynthia. *A Refreshing Look at Renewable Energy with Max Axiom, Super Scientist*. Capstone Press, 2020.

Renewable Energy Sources: Wave, Geothermal and Biomass. Baby Professor Books, 2017.

Websites

Find out more about biomass energy at:
www.ducksters.com/science/environment/biomass_energy.php

The Energy Information Agency (EIA) has information on renewable energy. For more information on biomass, visit:
www.eia.gov/energyexplained/index.php?page=biomass_home

A simple introduction to biomass energy can be found at:
www.real-world-physics-problems.com/biomass-energy-for-kids.html

Read up about algae and how it is changing biomass energy at:
www.renewableenergyfocus.com/view/931/biomass-from-algae

Publisher's note to educators and parents:
All the websites featured above have been carefully reviewed to ensure that they are suitable for students. However, many websites change often, and we cannot guarantee that a site's future contents will continue to meet our high standards of educational value. Please be advised that students should be closely monitored whenever they access the Internet.

Index

agriculture 9, 10, 11, 15, 17, 18, 19, 23, 27, 28, 29, 37, 41
algae 34–35, 37, 45
anaerobic digesters 10–11, 14, 15, 24, 31, 33
Asia 28–29
Australia 9, 15, 21, 33, 43, 45
aviation 20–21, 23, 43, 44, 45
Azolla 36–37

biodiesel 7, 16, 23, 25, 31, 34, 35
biofuels 5, 12, 16–17, 18–19, 20–21, 23, 24, 25, 26, 27, 31, 33, 34, 35, 36, 37, 40–43, 44, 45
biogas 5, 7, 10, 11, 12, 14, 22, 27, 28, 31, 33, 42, 43

carbon dioxide capture 12, 15, 32, 35, 37, 44
China 29
climate change 4–5, 37, 45
cocoa 41
coffee 40, 41
coppicing 19, 40
costs 9, 17, 25, 29, 35, 37, 40, 41
crops 5, 7, 16, 17, 18–19, 20, 21, 23, 24, 25, 26, 27, 28, 29, 30, 31, 32, 33, 40, 45

environmental impact 5, 8, 10, 12–13, 15, 18–19, 27
ethanol 7, 16, 17, 18, 20, 23, 25, 26, 27, 32, 33, 40, 43, 45
Europe 14–15, 22, 23, 24, 30–31, 32

farming 9, 10, 11, 15, 17, 18, 19, 23, 28, 29, 37, 41
fossil fuels 4–5, 12, 13, 14, 23, 25, 27, 45

gasification 39
global warming 37, 45
government support 25, 28, 42, 45
greenhouse gases 4, 5, 13, 19, 20, 25, 26, 32, 39, 43

heating 5, 6, 7, 8, 9, 11, 12, 14, 15, 23, 24, 25, 30, 32, 39, 41, 42, 43
hydrogen 38–39

India 22, 29
industrial uses 23
industrial waste 5, 20, 21

Kenya 22

landfill sites 10, 11, 24, 25
Latin America 26–27, 45

methane 7, 10, 11, 13, 15, 33, 44

plantations 8, 27
pollution 12, 13
power plants 4, 5, 7, 9, 12, 14, 22, 23, 24, 26, 27, 29, 30, 31, 32, 33

regulations 13

sawmills 8–9
storage 29, 33

transportation 16, 17, 18, 20–21, 25, 26, 31, 34–35, 37, 38–39, 42, 43, 44, 45

United Kingdom 14–15, 20, 21, 30, 31, 32
United States 13, 14, 16, 17, 18, 19, 20, 21, 23, 24–25, 30, 33, 43, 45

water heating 5, 7, 11, 15, 23
wood waste 6, 7, 8–9, 12, 13, 14–15, 19, 22, 23, 24, 27, 28, 30, 31, 32, 39

About the Author

Robyn Hardyman has written hundreds of children's information books on just about every subject, including science, history, geography, and math. In writing this book she has learned even more about science and discovered that innovation is the key to our future.

48